YOUR KNOWLEDGE HAS VALUE

- We will publish your bachelor's and master's thesis, essays and papers

- Your own eBook and book - sold worldwide in all relevant shops

- Earn money with each sale

Upload your text at www.GRIN.com and publish for free

Wind Tunnels. A detailed approach to analysing and comparing wind tunnel theory

Abdusselam Šabić

Bibliographic information published by the German National Library:

The German National Library lists this publication in the National Bibliography; detailed bibliographic data are available on the Internet at http://dnb.dnb.de.

ISBN: 9783346707925
This book is also available as an ebook.

Print and binding: Books on Demand GmbH, Norderstedt, Germany
Printed on acid-free paper from responsible sources.

The present work has been carefully prepared. Nevertheless, authors and publishers do not incur liability for the correctness of information, notes, links and advice as well as any printing errors.

GRIN web shop: https://www.grin.com/document/1264786

Table of Contents

Introduction

Wind tunnels are a great way of obtaining precise and accurate data on an aerofoil or model aircraft. They are crucial because they help engineers save a lot on resources; rather than having to make a full prototype by "winging it", as they say, with a wind tunnel, it is possible to test a collection of low-cost models prior to the prototyping stage in order to find the best configuration for a design.

As such, this assignment will cover the operation of wind tunnels from multiple points of view, looking at the various uses of a wind tunnel both from theoretical and practical points of view.

Task 1 – wind tunnel types

The type of wind tunnel used depends on the experimental application. The common types are subsonic, transonic, supersonic, hypersonic, and hypervelocity.

"Wind tunnel research has a wide range of applications, from normal aircraft testing to basic study on the boundary layer. At multiple locations on the model, measurements of air pressure and other variables provide information about how the entire wind load is distributed. Aerodynamic research in wind tunnels have proved extremely lucrative instruments for solving design difficulties in autos, boats, railways, bridges, and building structures, in addition to aviation and spacecraft." (Encyclopaedia Britannica, 2018)

Subsonic and transonic wind tunnels fall into the category of low speed wind tunnels. With a 100 horsepower variable frequency motor, "the low speed wind tunnel is a closed-circuit, continuous flow type. The flow area of the test portion is 0.6 x 0.9 metres (2 x 3 feet). A maximum flow velocity of 50 metres per second is possible in the tunnel. A 6-component force balance, a multiplexed data collecting system, and smoke visualisation equipment are also included in the tunnel." A high-pressure air supply is also offered as an option. (arc.uta.edu, 2014)

Subsonic wind tunnel

An aerodynamically optimised effusor (cone) draws air into the tunnel and accelerates it in a linear fashion. It then passes via a grille, a diffuser, and lastly a variable-speed axial fan to reach the working component. The grille protects the fan from stray objects. The air leaves the fan, passes through a silencer, and then returns to the outside world. A separate control and instrumentation unit, which also supplies electrical power to other instruments, controls the air velocity in the working component.

The tunnel's working space is a square with a transparent floor, sides, and roof. The sides can be taken off for access into the testing space. Wind tunnel models are supported by a particular region on the floor and each side panel. A protractor and a model holder are included with the wind tunnel to support and precisely modify the angle of any model installed.

The wind tunnel comes with two Pitot-static tubes. One is placed near the working section's intake and monitors the air flow rate. The second Pitot-static tube is coupled to a two-axis traverse that permits measurements to be taken both horizontally (fore and aft) and vertically across the working area.

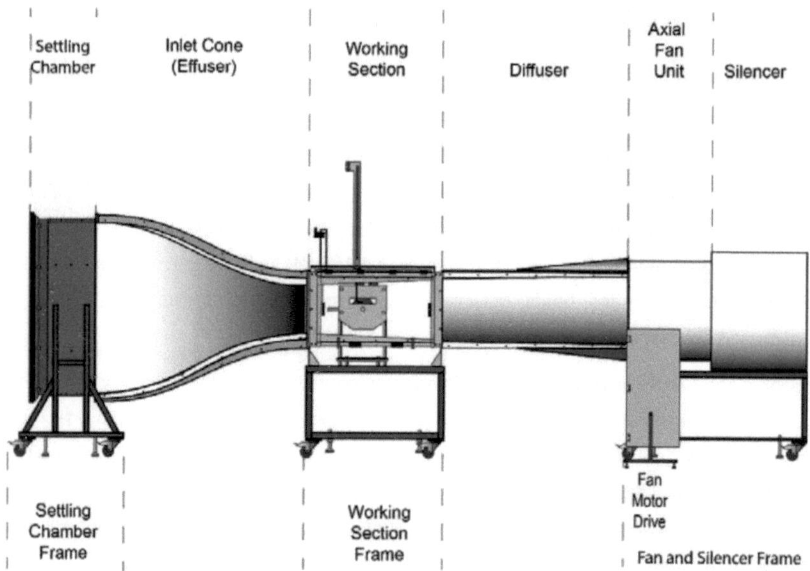

Figure 1 Subsonic wind tunnel cross-section diagram (TecQuipment LTD., 2021)

The air stream is upheld by a metal casing. For simple compactness, the edge has lockable castors. The coordinated hardware has electronic sensors that might be associated with TecQuipment's "Flexible Data Acquisition System" ("VDAS®", which is given). On an appropriate PC, "VDAS®" offers precise constant data gathering, observing, display, calculation, and outlining of every relevant boundary.

Typical uses of subsonic wind tunnels include:

- "Airflow past bluff and streamlined bodies with velocity and pressure observations in the wake"
- "Analysis of boundary layer propagation"
- "Aspect ratio's effect on performance of aerofoils"
- "Distribution of pressure around a cylindrical object under sub-critical or super-critical flow conditions"
- "Study of model characteristics that involve the basic measurement of drag and lift forces"
- "Analysing the characteristics of 3D aerofoils including drag, lift, and pitching moment forces"
- "Analysis of pressure distribution around an aerofoil model for deriving the lift force, and comparing the value obtained with direct lift measurements"

- "Measuring the drag force on a bluff body normal to the airflow"
- "Visualisation of airflow" (TecQuipment LTD., 2021)

Transonic wind tunnel

The "Transonic Wind Tunnel Göttingen" (TWG) is a closed return tunnel capable of airflow speeds ranging from subsonic to supersonic. The way it achieves this is due to its test section, which can be changed to fit one of three configurations as shown in the diagram below. The perforated test section is utilised for the transonic range of airflow.

Each of the three sections is 1x1x4.5 metres. The transonic test section has flexible lower and upper walls, which allows a 2D adaptation to the flow field. The wall interference is minimised compared to typical test sections, and/or bigger models in the range 0.3 M 0.9 may be employed. The adaptation is performed using wall pressure distribution data and a single-step technique based on Cauchy's integral formula for aerofoil (2D) tests and the Wedemeyer-Lamarche method for 3D models, respectively. Small residual wall interferences are estimated using Green's integral formula and utilised for final rectification in 3D model testing.

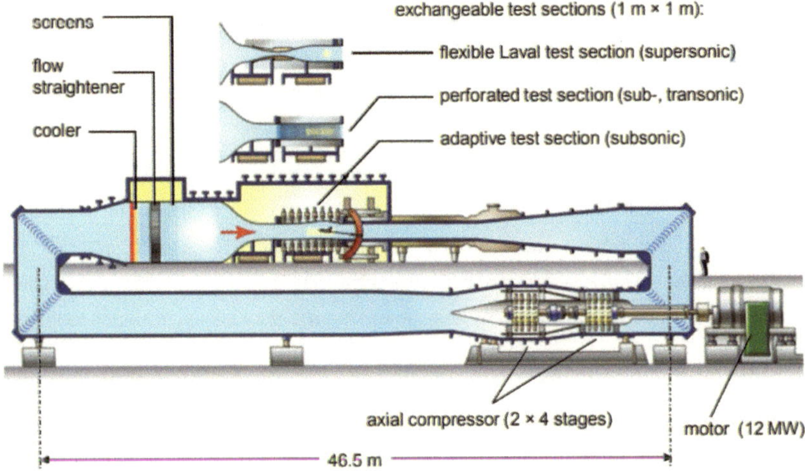

Figure 2 TWG wind tunnel (DNW, 2021)

Typical uses of transonic wind tunnels include:

- High precision tuning of Mach no. in increments of 0.001
- Continuous sweep measurement of pitch
- Direct simulation of sideslip angle
- Suction and blowing of heated air through a model or tunnel walls

- Simulation of air intake
- Drag analysis by duct flow measurement
- Free and forced pitch/heave oscillations and flutter simulation of 2D profiles and half models (DNW, 2021)

Advantages of low-speed wind tunnels

Open return wind tunnel

- Low construction cost.
- "Superior propulsion and smoke visualisation design. In an open tunnel, there is no build-up of exhaust products". (grc.nasa.gov, 2021)

Closed return wind tunnel

- For a given speed, the power required is smaller.
- Within the circuit, particulate debris can be confined.
- The amount of noise is greatly reduced.
- Wind tunnel flow is unaffected by laboratory air movement (air vents, doors, windows, and so on).
- There is no laboratory dust in the air that enters the test portion.
- Model failure does not cause as much harm to fan blades. (AEROLAB.com, 2021)

Disadvantages of low-speed wind tunnels

Open return wind tunnel

- "In the test portion, poor flow quality is possible. Extensive screens or flow straighteners may be required to turn the corner into the bell-mouth. The tunnel should also be situated away from any items in the room that cause asymmetries to the bell-mouth (walls, desks, people, etc.). Winds and weather have an impact on tunnels that are open to the atmosphere."
- "High operating costs due to the fact that the fan must continuously accelerate flow through the tunnel".
- "Generates high amounts of noise". (grc.nasa.gov, 2021)

Closed return wind tunnel

- "For a particular test section size, the cost is usually three times higher".
- "When dealing with combustion engines, the air supply is recycled, which might be a problem".
- "The footprint is significantly greater, necessitating more total area".

- "When used for a long time, the rising air temperature might become a problem". (AEROLAB.com, 2021)

Supersonic wind tunnel

A supersonic wind tunnel contains a "remote-control device that regulates a high-capacity vacuum pump is housed in an instrument frame (provided)". The pump produces low pressure downstream of the working area to pull air into the wind tunnel. The operator can reduce the air flow through the working component by employing a bypass duct with a hand-operated valve without impacting the main air flow quality. This is necessary for subsonic testing as well as beginning and stopping.

The working component of the wind tunnel is a convergent-divergent nozzle with a removable top half ('liner'). The shape of the liner determines the maximum air velocity at the diverging zone of the working section. There are three different shaped liners provided.

In the instrument frame, a 'mimic' panel and multi-pressure display unit are linked to pressure tappings along the working portion. The display device shows the pressures at the tappings. The display includes pressure sensors calibrated to monitor pressures relative to the atmosphere. One of the models' forces are also displayed.

An analogue pressure gauge is used to measure and display the pump's suction (tunnel reference pressure). This pressure line also connects to the multi-pressure display for data collection.

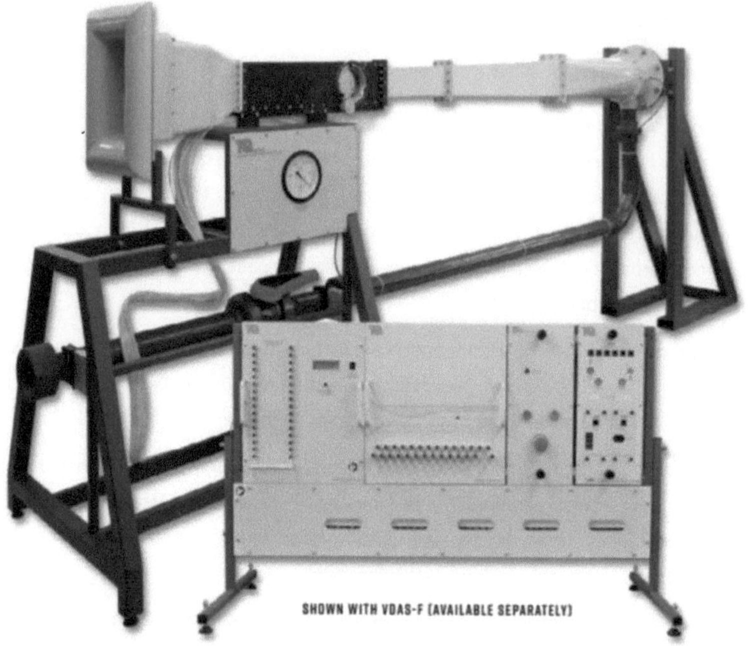

Figure 3 TecQuipment supersonic wind tunnel (TecQuipment, 2021)

Typical uses of supersonic wind tunnels include:

- "Pressure distribution with subsonic and supersonic air flows along a convergent/divergent (Laval) nozzle"
- "Real and theoretical area ratios of a nozzle at supersonic air speeds"
- "Comparison of theoretical and actual pressure distributions (Mach numbers)"
- "Pressures at different angles of incidence around a two-dimensional model under subsonic and supersonic flow circumstances"
- "Lift coefficients for supersonic flow aerodynamic models"
- "Shock waves and expansion patterns in supersonic flow around a two-dimensional model" (TecQuipment, 2021)

Advantages of supersonic wind tunnels

- "Mach capability is high. Tunnel "start-up" is simple and building and running expenses are low."
- "Excellent propulsion and smoke visualisation design. In an open tunnel, there is no build-up of exhaust products."

- "Faster starts result in smaller loads on the model at start-up."

Disadvantages of supersonic wind tunnels

- Faster (and typically more costly) instrumentation is required for shorter test periods.
- Pressure regulator valves are required.
- It's a noisy process. (grc.nasa.gov, 2021)

1. Introduction

1.1 Experiment 1 – Wind tunnel calibration

The aim of this experiment is to understand the operation of an open return subsonic wind tunnel by learning how to calibrate the instruments on the wind tunnel. The instrumentation calibrated included a pressure transducer (pitot probe) and manometer.

Two sets of readings were taken: reference velocity variation with manometer height and reference velocity variation with probe height.

1.2 Experiment 2 – NACA 2412 aerofoil with variable flap

The second experiment involves the NACA 2412 3D aerofoil profile. The aim of this experiment is to "determine the effect that the flap deflection angle has on the aerofoil's aerodynamic characteristics".

2. Objectives

2.1 Experiment 1 – Wind tunnel calibration

- "Investigate the velocity at the test section inlet"
- "Familiarise ourselves with wind tunnel operation"
- "Examine the velocity uniformity"
- "Check the boundary layer height"

2.2 Experiment 2 – NACA 2412 aerofoil with variable flap

- "Determine the effect of flap deflection angle on the aerofoil's aerodynamic characteristics".

3. Theory

3.1 Experiment 1

Measuring velocity using a pitot probe

Visualising flow patterns and measuring pressure at a given place in the flow field and computing the accompanying air speed are two significant applications of wind tunnels. The equation connects the fluid's speed at a given position to its mass density as well as the pressures at that same location in the flow field. For an incompressible fluid with steady flow for which viscosity is negligible, the equation of flow velocity is given by:

$$v = \sqrt{\frac{2(P_0 - P)}{\rho}}$$

Where v = fluid speed, P_0 = total (stagnation) pressure, and P = static pressure.

This can be rewritten as:

$$v = \sqrt{\frac{2\Delta P_l}{\rho_a}}$$

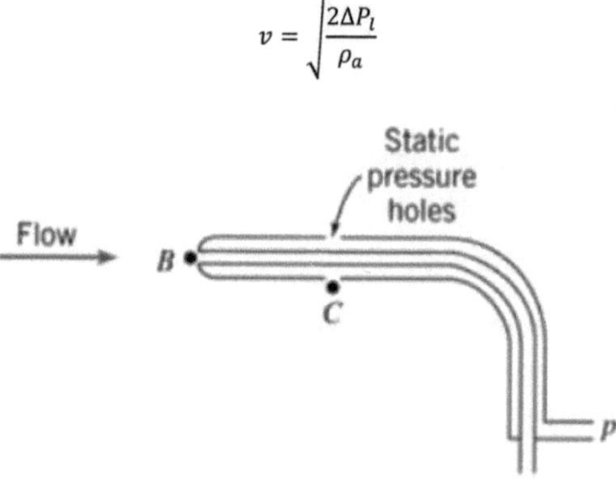

Figure 4 Pitot static tube (vlab.amrita.edu, 2015)

Bernoulli's equation for the steady flow of an incompressible and inviscid fluid down a streamline is the source of this equation. Integrating Euler's equations along a streamline usually yields Bernoulli's equation. It should be remembered that "Euler's equations are a specific case of the Navier-Stokes equations in which the fluid's viscosity is ignored. When Newton's 2nd law is applied to a fluid whose shear deformation follows Newton's law of viscosity, the Navier-Stokes equations are produced". (vlab.amrita.edu, 2015)

Measuring pressure distribution using a manometer

A manometer is used for measuring pressure differentials between 2 or more points. A simple manometer has 2 pressure ports. Each pressure port has a pressure exerted on it. The hydrostatic equation that links the pressure difference between the two ends is given by:

$$P_1 = P_2 + \rho_f g \Delta h, \qquad \rho_f = manometer\ fluid\ density$$

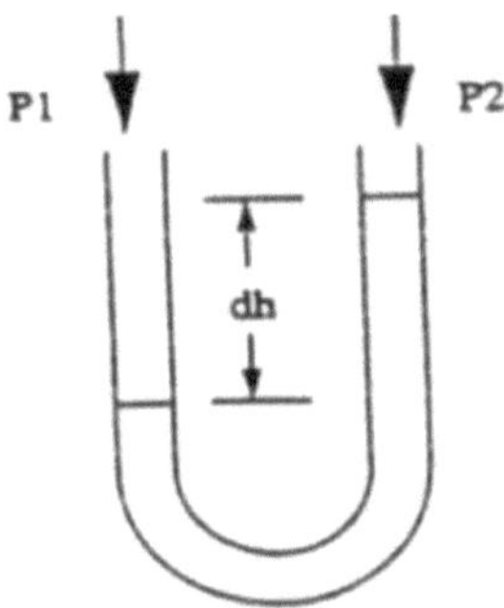

Figure 5: Simple manometer (Dr. Hui Hu, 2021)

A wind tunnel has a multi-tube manometer such as the one shown in the diagram below.

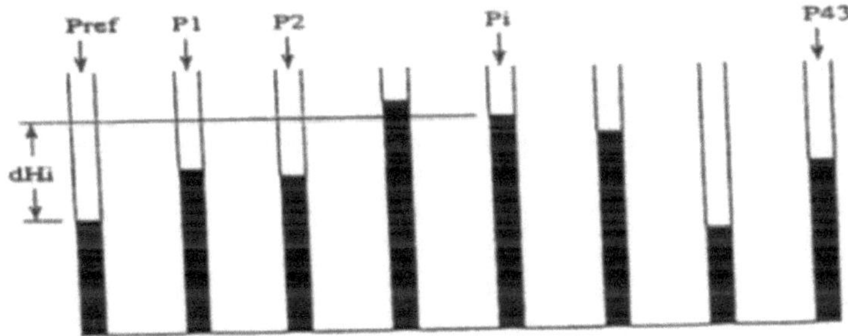

Figure 6: Multi tube manometer (Dr. Hui Hu, 2021)

Each subsequent tube paired with the tube after it can be considered as a separate manometer. Even though the fluid is shared throughout all the tubes, with a big enough reservoir, tube interference may be ignored. The vertical height difference between the fluid levels of the reference and the ith tubes are denoted by Δh_i. Therefore, using the hydrostatic equation, the following equation is derived:

$$P_{ref} = P_i + \rho_f g \Delta h$$

The absolute reference pressure is needed to get the absolute pressure at the ith tube. Because the pressure is usually ambient, or room pressure, a barometer can be used to find P_{ref}. (Dr. Hui Hu, 2021)

Boundary layer theory

As an object passes through a fluid or as a fluid flows past an item, the molecules of the fluid around the body are disturbed and move around the object. Aerodynamic forces are formed between the fluid and the object. The quantity of these forces is governed by the item's shape, speed, and the mass of the fluid travelling through the object, as well as two other crucial fluid characteristics: viscosity and compressibility. To properly mimic these effects, aerospace engineers use similarity parameters, which are ratios of these effects to other forces involved in the issue.

The viscosity of the fluid has an intricate effect on aerodynamic forces. "As the fluid passes over the body, the molecules closest to the surface cling to it. When molecules close above the surface collide with molecules that adhere to the surface, they slow down. These molecules, in turn, cause the flow right above them to slow down. The more one walks away from the item surface, the less collisions are affected. This results in a thin layer of fluid at the surface with a velocity that changes from zero at the surface to the free stream value further from it. Because it occurs on the fluid's border, engineers named this layer the boundary layer." (grc.nasa.gov, 2021)

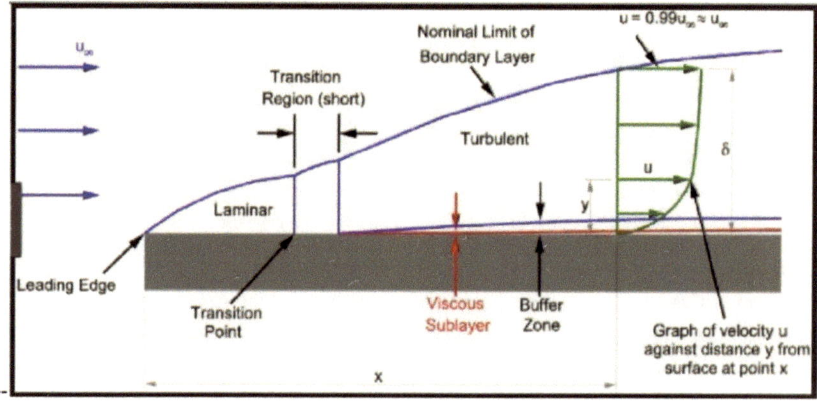

Figure 7 Boundary layer diagram (empoweringpumps.com, 2021)

"Many aspects of aerodynamics, such as wing stall, skin friction drag, and heat transfer during high-speed flight, rely heavily on the specifics of the flow within the boundary layer". The flow-wise velocity fluctuation from free stream to surface is seen in the diagram. In reality, the impacts are 3-dimensional. Because of the conservation of mass in three dimensions, a change in velocity in one direction causes a change in velocity in the other.

The flow above the surface is displaced or moved by a little component of velocity perpendicular to the surface. The amount of displacement may be utilised to establish the boundary layer thickness. The Reynolds number, which is "the ratio of inertial forces to viscous forces", determines the thickness of the displacement. The Reynolds number (Re) is calculated using the following formula:

$$Re = \frac{v\rho L}{\mu}$$

Where v = velocity, ρ = density, L = characteristic length, and μ = viscosity coefficient.

"Depending on the Reynolds number, boundary layers can be either laminar or turbulent. As seen on the left side of the diagram above, the boundary layer is laminar for lower Reynolds numbers, and the stream-wise velocity varies evenly as one advances away from the wall. The boundary layer becomes turbulent at increasing Reynolds numbers, and the stream-wise velocity is defined by unstable (changing over time) swirling flows inside the boundary layer. The external flow reacts to the boundary layer's edge in the same way it would to an object's actual surface." (grc.nasa.gov, 2021)

Force coefficients from distribution of pressure

Integrating the surface pressure coefficient distribution yields the drag, lift, and pitching moment coefficients. The lift force is the force acting on the aerofoil section perpendicular to the mean flow direction. When the nose is up, the pitch moment is a positive instant that occurs at the quarter chord point. Aerodynamic measurements are collected in the centre of the aerofoil section, assuming 2-dimensional flow. In this case, considering the force and moment per unit span is more realistic. The coefficients of section lift, pressure drag, and moment are defined as follows:

$$C_L = \frac{L}{\frac{1}{2}\rho U^2{}_\infty c} \quad , C_d = \frac{D}{\frac{1}{2}\rho U^2{}_\infty c} \quad , C_m = \frac{M}{\frac{1}{2}\rho U^2{}_\infty c}$$

Where L = lift force per unit span, D = drag force per unit span, M = pitching moment per unit span, and c = chord length. (vlab.amrita.edu, 2015)

3.2 Experiment 2

Aerofoil theory - forces on an aerofoil

Lift and drag forces are connected to stress distributions on a body via integration. Consider the stress that the aerofoil in figure is under. There is a shear stress distribution as well as a pressure distribution. Select a differential region as illustrated in the image below to connect stress to force. The pressure force has a magnitude of dF$_p$ = pdA, whereas the viscous force has a value of dF$_v$ = dA.

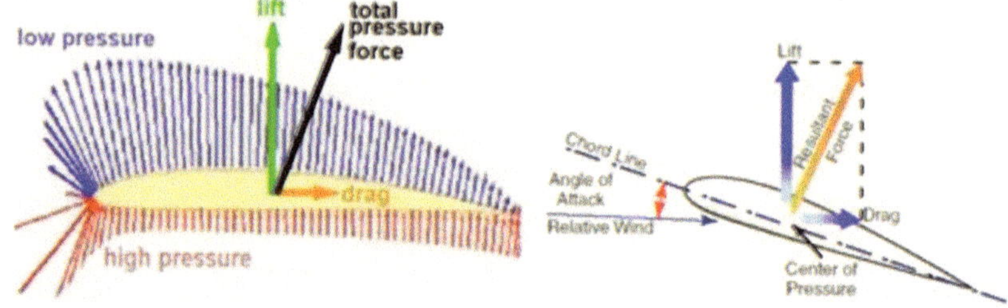

Figure 8: Pressure difference between the top and bottom of an aerofoil (left), and free-body diagram showing the forces on an aerofoil (right) (vlab.amrita.edu, 2015)

"The differential lift force is perpendicular to the direction of the free stream".

$$dF_L = -pdAsin\theta - \Gamma dAcos\theta$$

"The differential drag is parallel to the free-stream direction", hence:

$$dF_D = -pdAcos\theta + \Gamma dAsin\theta$$

Integrating over the aerofoil's surface results in a mathematical expression for the lift force and drag force, respectively.

$$F_L = \int (-psin\theta - \Gamma cos\theta)dA$$

$$F_D = \int (-pcos\theta + \Gamma sin\theta)dA$$

The equations for F$_L$ and F$_D$ defined above show that "lift and drag are related to pressure distribution by integration".

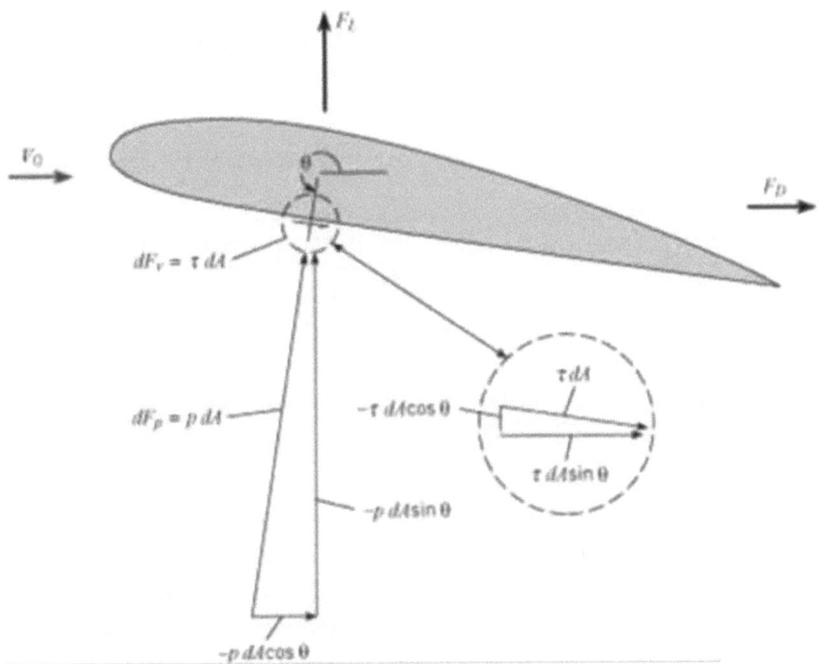

Figure 9: Diagram showing the viscous forces and pressure acting on a differential area element (vlab.amrita.edu, 2015)

Drag force equation

The drag force is found by using the lift equation as follows:

$$F_D = \frac{1}{2}\rho v^2 C_D A$$

Where C_D = drag coefficient, A = surface area of aerofoil, ρ = fluid density, v = free-stream velocity relative to the aerofoil.

Therefore, both lift and drag forces are dependent on 4 variables:

- The object's shape (characterised by C_L and C_D)
- The object's size (characterised by A)
- The ambient fluid's density
- The square of fluid velocity

The area of reference A is determined by the body type. The silhouetted area that would be seen by a person looking at the body from the direction of flow is one typical reference area, known as projected area and denoted by the sign Ap. The planform area of an aircraft wing serves as the reference area. This is the region visible from above while looking at the wing. The drag force associated with a specific body form is characterised by the C_D parameter. Experiments are commonly used to determine the worth of a C_D. A force balance in a wind tunnel may be used to determine the drag force.

The drag force equation can be rearranged to obtain an expression for the drag coefficient:

$$C_D = \frac{F_D}{\frac{1}{2}\rho v^2 A}$$

The air speed in the wind tunnel "V_0" may be monitored using a pitot-static tube or equivalent instrument, and air density can be computed using the ideal gas law and recorded temperature and pressure data. A pitot-static tube or similar device may be used to measure air speed in the wind tunnel "V_0", and air density can be calculated using the ideal gas law and recorded temperature and pressure data.

The circulation in the flow created by the aerofoil causes lift. Consider the flow of a non-viscous, incompressible ideal flow past an aerofoil as shown in the diagram below. Lift and drag are zero in this case, as they are in irrotational flow past a cylinder. On the bottom side, near the leading edge, there is a stagnation point, and on the top side, near the trailing edge of the foil, there is another. The flow pattern around the upstream half of the aerofoil in a reasonable flow (viscous fluid) seems reasonable. However, the flow pattern in the trailing edge area is not possible. Fluid must flow from the bottom side around the trailing edge and then toward the stagnation point if there is a stagnation point on the top side of the foil.

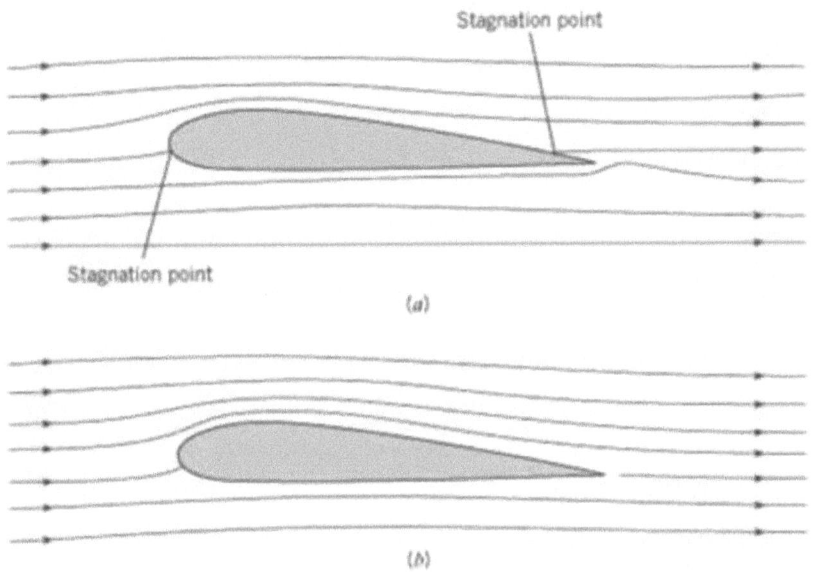

Figure 10: Flow patterns around an aerofoil (vlab.amrita.edu, 2015)

Lift force equation

To reconcile theory with the observed event, it was proposed that a circulation around the aerofoil be produced in precisely the correct quantity to transfer the downstream stagnation point all the way back to the trailing edge of the aerofoil, allowing the flow to exit the aerofoil smoothly. The name given to this is Kutta condition. When analysing the flow pattern and pressure distribution, as well as the lift on a two-dimensional aerofoil section, this basic assumption about the size of the circulation yields extremely excellent agreement between theory and experiment (no end effects).

The magnitude of the circulation required to keep the rear stagnation point at the trailing edge of a symmetric aerofoil with a small angle of attack of a symmetric aerofoil with a small angle of attack of a symmetric aerofoil with a small angle of attack of a symmetric aerofoil is given by:

$$\Gamma = \pi c v \alpha$$

Where Γ is the circulation, c = aerofoil chord length, and α = the chord's angle of attack relative to the free stream.

The lift per unit length for an infinitely long wing is given by:

$$\frac{F_L}{L} = \rho v \Gamma$$

The length planform area is L, so the lift on the aerofoil segment is given by:

$$F_L = \rho v^2 \pi c L \alpha$$

Thus, the lift force is found by using the lift equation as follows:

$$F_L = \frac{1}{2}\rho v^2 C_L A$$

Where C_L = lift coefficient, A = surface area of aerofoil, ρ = fluid density, v = free-stream velocity relative to the aerofoil.

This equation can be rearranged to obtain an expression for the lift coefficient:

$$C_L = \frac{F_L}{\frac{1}{2}\rho v^2 A}$$

(vlab.amrita.edu, 2015)

Pitching moment equation

The net lift and drag force operates on the aerofoil's centre of pressure. The centre of pressure, on the other hand, is not a fixed place and will change when the angle of attack of the aerofoil changes. Because the centre of pressure is not fixed, it is not a practical site to describe the resulting forces operating on the aerofoil. The usage of a point designated at the aerofoil quarter chord is a standard convention. One fourth of the way along the chord from the leading edge is this point. To create a force balance, a moment must be applied to the resulting lift and drag force from the centre of pressure to the quarter chord.

To attain static equilibrium, a "pitching moment equal to the lift force multiplied by the moment arm between the quarter chord and the centre of pressure is applied" ("Here the component of the shear force that would contribute to the total pitching moment is neglected as it is negligibly small relative to the lift component"). The resultant aerodynamic force on the aerofoil may thus be described as a "lift and drag force operating at the quarter chord, plus a balancing pitching moment". (AeroToolbox.com, 2017)

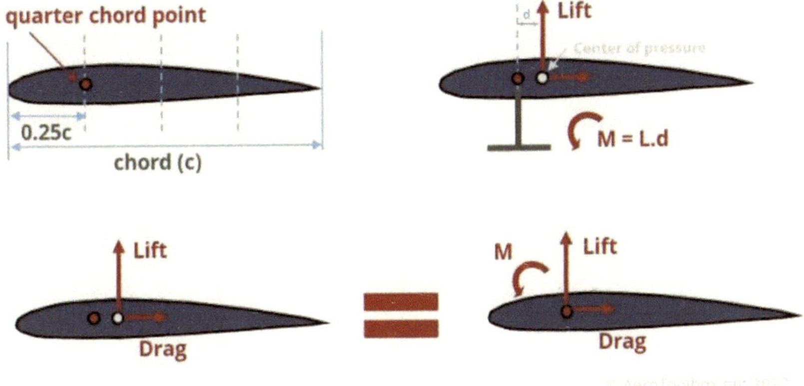

Figure 11: Quarter chord location (top left), and resolving the force vectors to their respective components and pitching moment (AeroToolbox.com, 2017)

The coefficient of pitching moment is given by:

$$C_M = \frac{M}{\frac{1}{2}\rho v^2 A c}$$

High lift devices

The main purpose of high lift devices is to allow for steeper approaches without increasing the aircraft speed, as well as reducing the take-off and landing distance. High lift devices are also used for reducing the speed at which the aircraft stalls, and for increasing the pilot's visibility of the runway. (Ria & Kartik, 2021) (aviatorsbuzz.com)

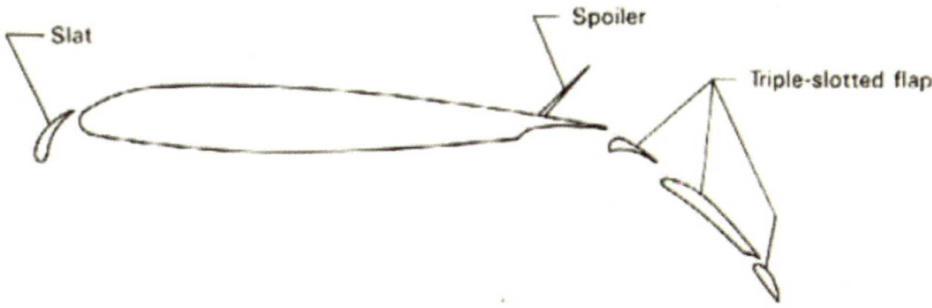

Figure 12: Aerofoil with triple-slotted flap, slat, and spoiler (Laurence K. Loftin, 2012) (history.nasa.gov)

The image above shows a wing section with a leading-edge slat and triple-slotted flap on the trailing edge. The trailing-edge flap divides into 3 sections as it deploys backward and downward. The flap's slots allow fluid to flow from the bottom to the upper surface.

The flow through the slots energises the top-surface boundary-layer flow, which is "negotiating a positive pressure gradient, and avoids separation and eventual loss of lift". The slot shapes' exact design is crucial, and it must be meticulously worked out in experimental tests. During cruise, both the leading and trailing-edge flaps are totally retracted and only deployed for landing and take-off.

Figure 13: Aerofoil with a Krueger flap and double-slotted flap (Laurence K. Loftin, 2012) (history.nasa.gov)

The image above shows a wing section with a Krueger flap on the leading edge and double-slotted flap on the trailing edge. The Krueger flap is less effective than the slat, but its mechanical design is presumably simpler. Where more forceful flow control is necessary, some aircraft use slats on the outboard section of the leading edge and Krueger flaps on the inboard portion.

The triple-slotted trailing-edge flap is more powerful, while the double-slotted trailing-edge flap is mechanically simpler and easier to execute. As a trailing-edge device, the basic single-slotted flap is frequently utilised. This flap is made up of a single unsegmented part that is moved backward and downward to deploy. Although it is less efficient than the other two types of trailing-edge devices mentioned, it is by far the most mechanically basic of the three, with the simplest aerodynamic shape. (Laurence K. Loftin, 2012) (history.nasa.gov)

4. Instrumentation used on the wind tunnel

Below is an image showing the control panel of the wind tunnel used for the calibration experiment. The right-hand side manometer displays the total pressure, and the left-hand side manometer displays the static pressure.

The list of instrumentation used for this experiment is:

- AF100 wind tunnel

- Three-component balance (AFA3) which includes:
    - Differential pressure transducer
    - Multi-tube manometer (AFA1)
- Pitot static probe
- AF103 NACA2412 variable flap aerofoil

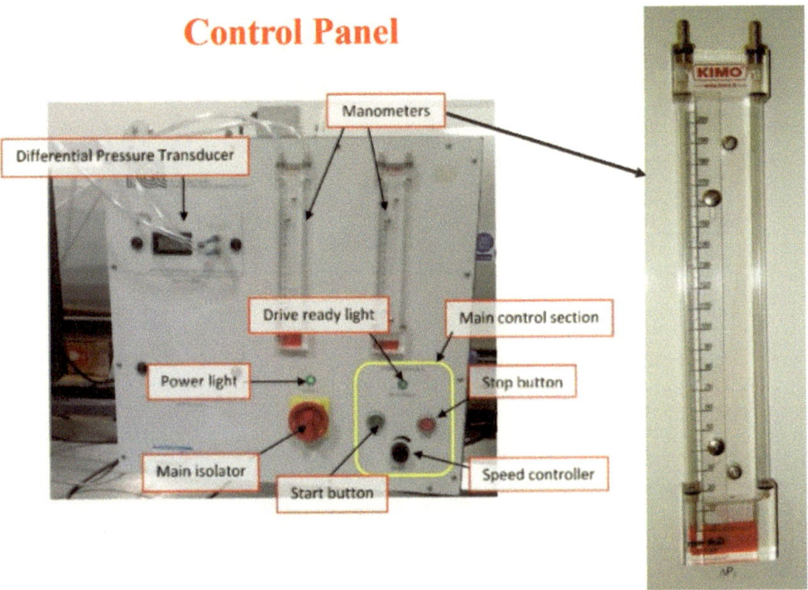

Figure 14: Control panel of the wind tunnel used for the first experiment, including the manometer (Sreeja Sreekumar PPT, 2021)

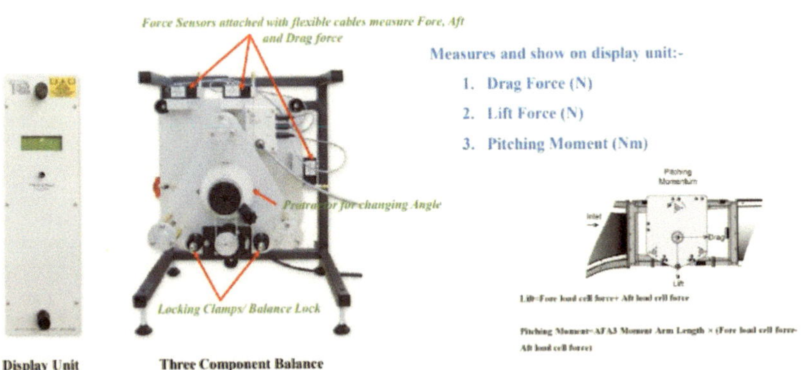

Figure 15: 3-component balance apparatus that calculates the lift and drag forces based on the tension in the cables (Sreeja Sreekumar PPT, 2021)

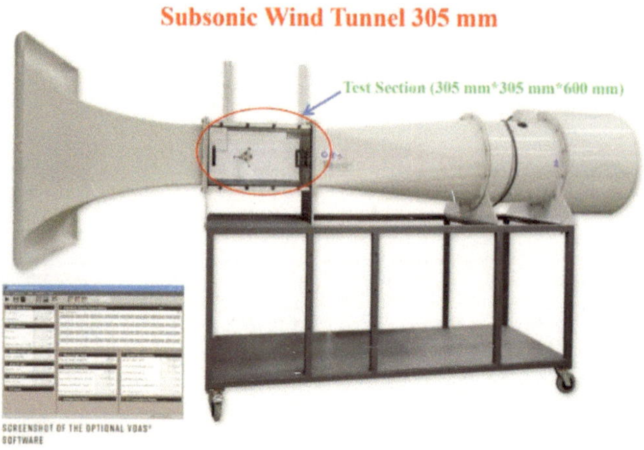

Figure 16: Wind tunnel used for this experiment (Sreeja Sreekumar PPT, 2021)

Figure 17: Pitot static tube (infoWERK, n.d.)

Figure 18: 150 mm chord NACA2412 variable flap aerofoil (infoWERK, n.d.)

5. Precautions and procedure

5.1 Precautions

- "Ensure that all components and test objects are tightly secured prior to starting the wind tunnel."
- "Ensure that all models are reviewed prior to being mounted in the wind tunnel"
- "Any loose objects might travel to the blow-housing and result in severe injury to personnel or severe damage to the equipment."
- "When starting the wind tunnel, if abnormal sounds are heard while the blower and motor are starting up, immediately press the red STOP button." "The wind tunnel should not be used again until the manufacturer is notified, and the problem is resolved." (eng.yale.edu, 2021)
- "If you lack the necessary skills or are unfamiliar with your own equipment, do not attempt to accomplish anything."
- "Allowing oneself to get sidetracked needlessly is a bad idea, and if you're with others, make sure they know what's going to happen."
- "Make sure there are no loose items or anything linked to the wind tunnel that might be displaced before operating it."
- "Never turn on the wind tunnel if you don't need it. Bring the flow to a halt after the test is done."
- "Never distract someone while they are working on the wind tunnel."
- "If your setup has components that spin at a high rate, be extra cautious." (gla.ac.uk, n.d.)

5.2 Procedure

5.2.1 Experiment 1 – reference velocity

1. "Make sure the Pitot tube is positioned at the front (near the intake). Make sure the mechanism permits the probe to go the whole length of the portion." "If not, unscrew the screw that holds the pitot tube to the pointer and make the necessary adjustments."
2. "Connect the wall tapping and pitot tube to the manometer ΔP_i, connecting the wall tapping to the right-hand side manometer and the pitot tube to the left-hand side manometer."
3. "Take readings of the barometric pressure and ambient air temperature after zeroing the manometers." "Place the pitot tube in the tunnel's centre (152.5 mm from the bottom)."
4. "Start the fan and set the speed to 22 m/s. Using the manometer, take a reading. Reduce the fan speed and check the manometer again." "Continue to collect readings on the manometer for a variety of fan speeds."
5. "Plot a graph of the averaged results".

Experiment 1 - Velocity profile

1. "Set the pitot tube to the middle position and the fan to a speed of roughly 5 m/s, 10 m/s, 15 m/s, 20 m/s, and 25 m/s (use the pitot-static tube data display)."
2. "Place the pitot on the test section floor. Take readings of the dynamic pressure and pitot position from the scale."
3. "Repeat the dynamic pressure reading by moving the pitot away from the wall by about 1 mm. Continue taking dynamic pressure readings every millimetre up to 15 mm, then every 20 mm up to the test section's centre line."
4. "Plot a graph of the height gained by the velocity in the test section (subtract the radius of the pitot tube for a true position relative to the tunnel floor)."
5. "The graph should indicate a lower velocity near the tunnel floor owing to the boundary layer, but a somewhat consistent velocity for the rest of the test segment."

5.2.2 Experiment 2 – NACA 2412 variable flap aerofoil

Starting the wind tunnel

1. "Power on the electrical isolator on the instrumentation control frame."
2. "Set the speed control to the lowest setting (totally anticlockwise)"
3. "Press the green START button"
4. "Gradually turn the speed control clockwise until the tunnel is running at the speed required for the experiment."

Inserting the aerofoil model

1. "Ensure that the wind tunnel's electrical supply is disconnected."
2. "Remove the side window opposite the 3-component balance."
3. "Ensure that any pitot static tubes in the test section are not in the way."
4. "Place the aerofoil model in the AFA3's collect such that the flattest side is on top. The load cells and connecting wires on the AFA3 are intension since the aerofoil is flying upside down. The pitch is positive, with the leading edge pointing downwards to the floor."
5. "Adjust the variable flap to align with the main section of the airfoil and the metal plate at the airfoil's support shaft end. The flap angle has now been set to zero."
6. "Set the AFA3's scale to zero and lock it in place. Rotate the aerofoil until the trailing edge is 2.6 mm higher than the support shaft's centre line. On the AFA3, secure the support shaft in place. The 2.6 mm offset compensates for the aerofoil's camber. The angle of the aerofoil is not zero."
7. "Refit the clear panel that was removed in step 2."

Wind tunnel test procedure

1. "Record the ambient temperature and pressure."
2. "Release the balance lock prior to commencing with the testing procedure."
3. "Start the wind tunnel and set the free-stream velocity for each team. Because of the ambient circumstances, the pressure difference utilised to determine the wind velocity changes."
4. "The readings from the Three component balance display unit for lift, drag, and pitching moments should be recorded."
5. "Adjust the angle of attack in 5° increments from -5° (355) to 10°."
6. "The aerofoil is flown upside down in the 3-component balance. This is quite typical and enables for more precise balance design."
7. "Take measurements of fore and aft drag forces at aerofoil angles ranging from -5° to 10° in increments of 5°. Take 3 readings for each angle of attack, since this will help to decrease mistakes caused by variations."
8. "Stop the wind tunnel, remove the transparent panel opposite the AFA3, and set the flap angle to 5 degrees."
9. "Repeat the experiment using a 10-degree flap angle."
10. "Tighten the balance lock after test."

6. Results and discussions

6.1 Experiment 1

6.1.1 Reference velocity

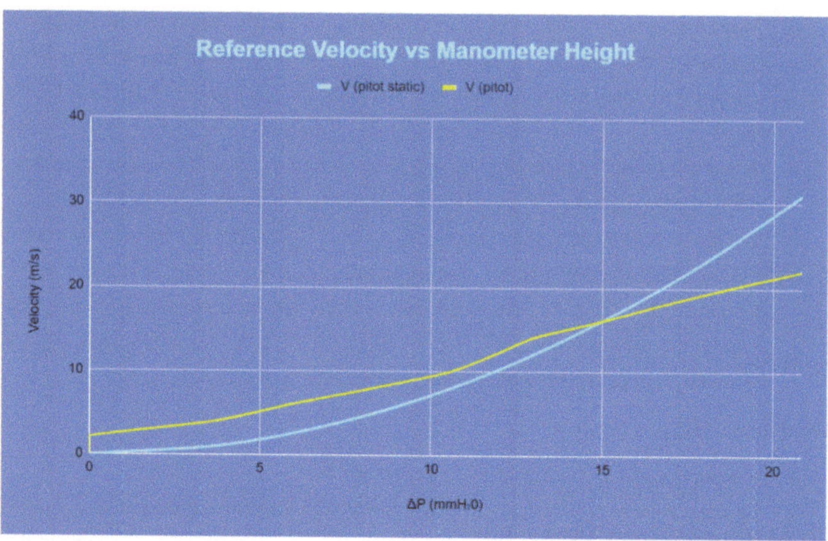

The reference velocity is the combination of pitot static and pitot velocities. The pitot static velocity is a range of values of the independent variable. Using the independent variable and the dependent variable (ΔP_l), static velocity is calculated.

The reference velocity is an indicator of the disparity between the inlet and exit velocities, and this is shown as 2 separate data plots in the graph above.

Example calculation:

$$\rho_a = \frac{P_a}{RT_a} = \frac{101325}{287(288)} = 1.22586$$

$$V_{(pitot)} = \sqrt{\frac{2 \cdot \Delta P_l}{\rho_a}} = \sqrt{\frac{2(870 \times 9.81 \times 31 \times 10^{-3})}{1.22586}} = 20.77 \; ms^{-1}$$

6.1.2 Velocity profile

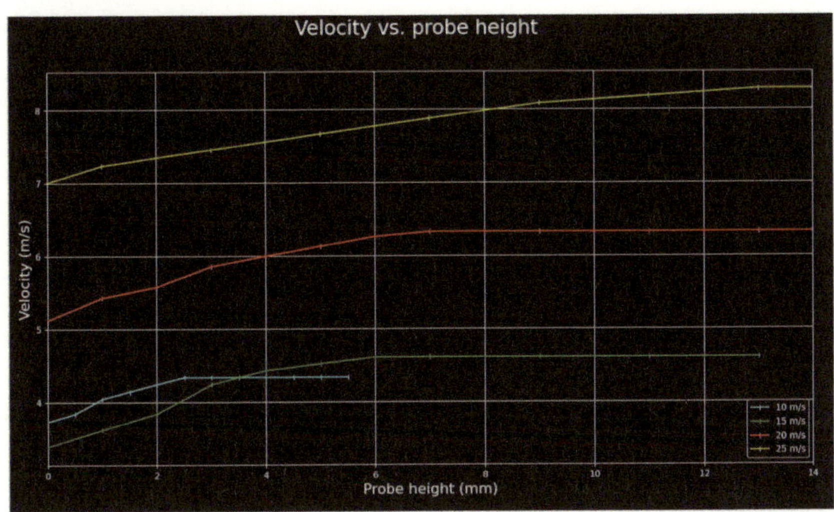

The graph above shows the 4 sets of readings for velocity at varying probe height for 4 speed ranges. The graph shows that as the probe height increases, the velocity also increases. This is due to a reduced flow velocity near the test section (tunnel) floor caused by the boundary layer with a quasi-constant velocity in the remaining parts of the test section.

6.2 Experiment 2

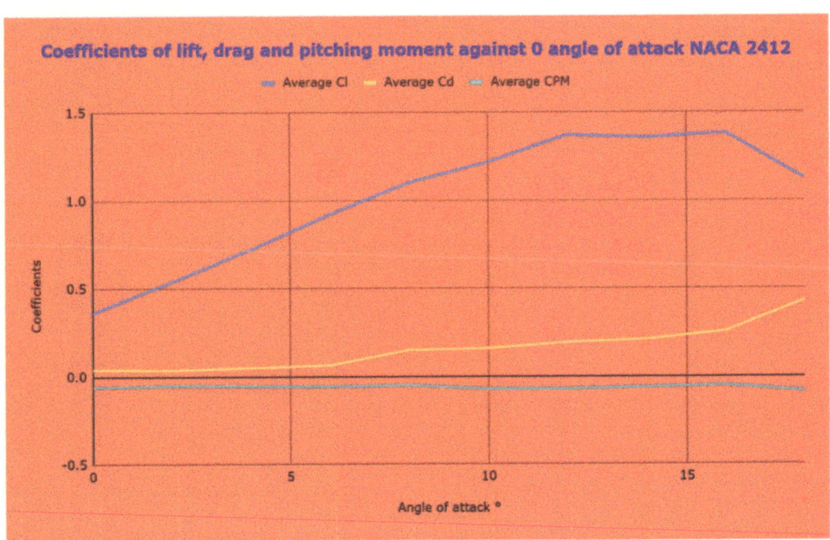

This graph shows the variation of the coefficients of lift, drag, and moment with angle of attack. The maximum lift coefficient arises somewhere around 12 degrees angle of attack, and this curve levels off until at 16 degrees angle of attack, which is where the aerofoil starts stalling. 12 degrees is the optimal angle of attack because C_d is lower here than at angles between 13-16, even though they have the same C_l.

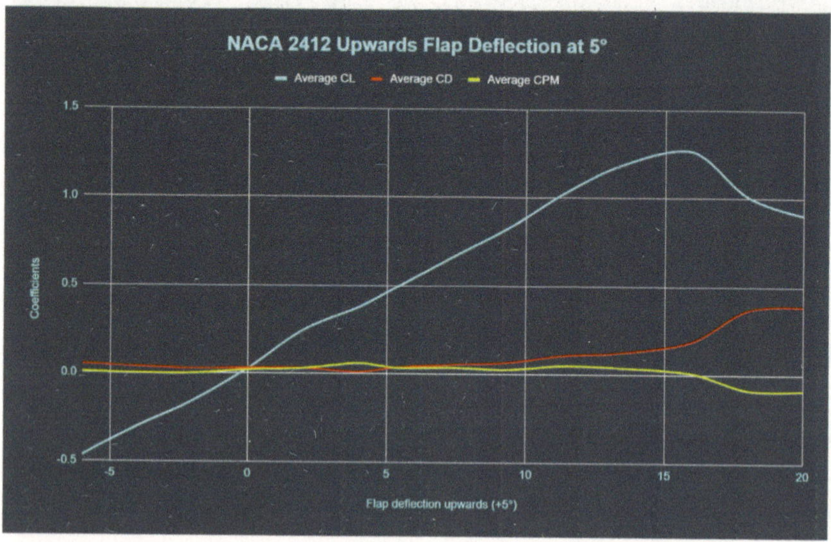

When the flap was deflected to 5 degrees, the lift produced by the aerofoil no longer levelled off at 12 degrees, which means that the optimal angle of attack with this flap angle is 16 degrees. However, an angle of 14 or 15 would be more feasible due to the lower drag coefficient there. The decrease in C_l for lower angles of attack is due to the overall effective decrease in upper camber area.

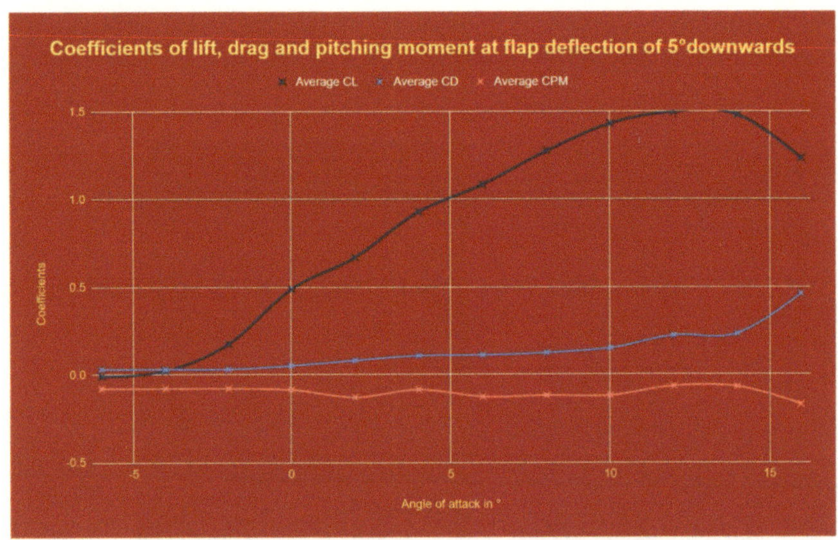

When the flap is deflected by 5 degrees in the opposite direction, it is noticeable that the total lift coefficient is greater for each angle, with the optimum angle being 13 degrees. The increase in C_l is due to the overall effective increase in upper camber area.

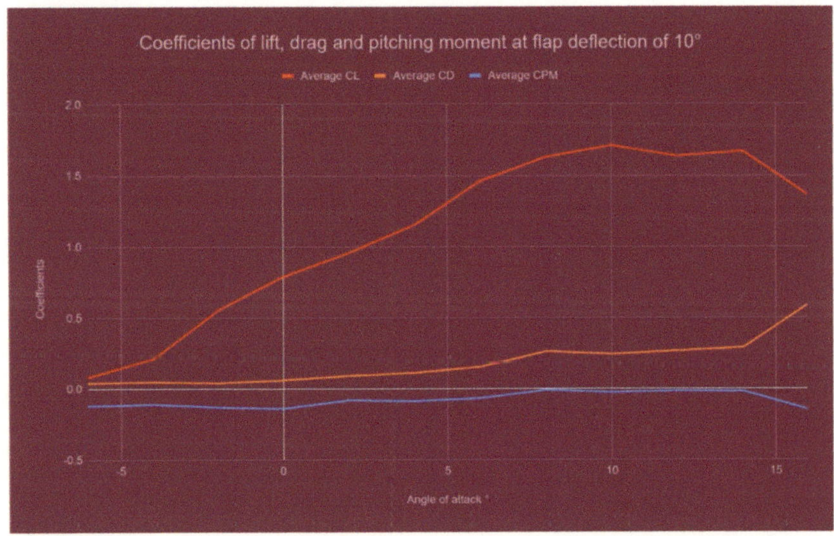

When the flap deflection angle is incremented by another 5 degrees, the overall C_l for each angle of attack increases again. The optimum angle of attack in this case is 10 degrees.

This experiment shows that as the flap angle is increased, the optimum angle of attack decreases because more lift is generated at lower angles of attack when the flap is extended.

Comparing the coefficients together

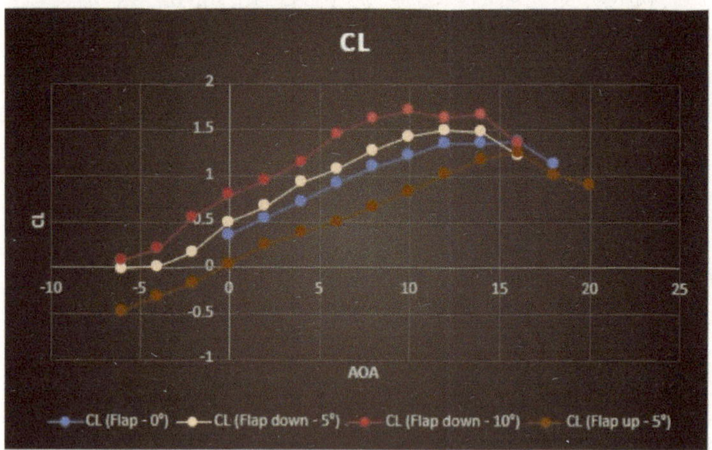

As mentioned above, the highest CL values were observed when the flap was deflected by 10 degrees downwards. Conversely, the lowest CL values were for an upward flap angle of 5 degrees, due to the fact that an upward flap deflection reduces the lift. This is useful when landing an aircraft, and the opposite is useful when taking off.

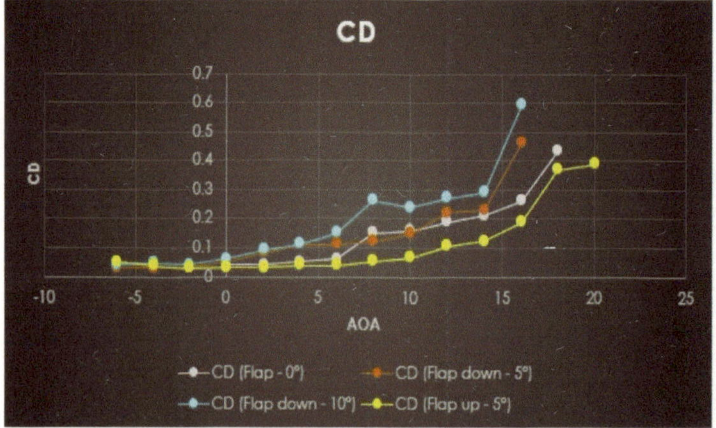

This graph shows that as the flap angle is increased, CD also increases. So, even though a flap angle of 10 degrees has the highest CL values, it also has the highest CD values. The same pattern is observed for an upward flap angle of 5 degrees.

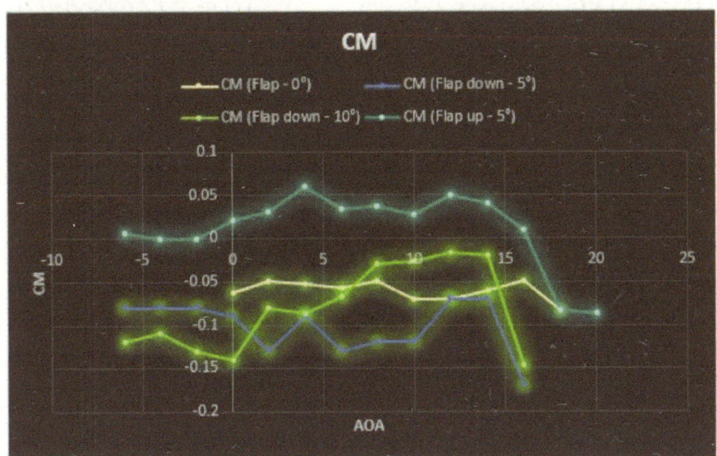

The highest pitching moment coefficients correspond to an upward flap angle of 5 degrees. This means that deflecting the flap upwards reduces the stability of the aerofoil. The opposite is observed as the flap is deflected lower, with a flap down angle of 10 degrees having the lowest pitching moment coefficients, meaning that it's the most stable configuration here.

7. Sources of error

Measurement errors can be classified as random or systematic, depending on how they were obtained (an instrument could cause a random error in one situation and a systematic error in another).

Random errors are stochastic variations (in either direction) in measured data produced by the accuracy limits of the measuring equipment. Random errors may be examined statistically and reduced by averaging across a large number of data points.

Systematic mistakes are errors that occur repeatedly and always point in the same direction. These faults are difficult to detect and cannot be quantified. A systematic inaccuracy can be found while calibrating against a standard, and the bias can be decreased by applying a correction or correction factor to compensate for the effect. Unlike random mistakes, systematic errors cannot be detected or reduced by increasing the number of observations.

Instrument resolution (random) - The precision of all instruments limits the ability to resolve slight measurement differences. A metre stick, for example, cannot distinguish distances to more than half of its lowest scale division (0.5 mm in this case). One of the most effective methods for obtaining more precise measurements is to use a null difference technique rather than directly measuring a quantity. The use of instrumentation to identify the difference between two identical values, one of which is well-known and configurable, is the basis for null or balancing techniques. By altering the adjustable reference amount, the disparity is reduced to zero. After balancing the two values, the unknown quantity's amount may be calculated by comparing it to a measurement standard. This method removes source instabilities, and the measuring equipment may be exceedingly sensitive without requiring a scale.

Zero offset error (systematic) - When using a micrometre calliper, electronic balance, or electrical metre, always verify the zero reading first. Re-zero the instrument if possible, or at the absolute least measure and record the zero offset so that data may be corrected later. It is also a good idea to check the zero reading during the experiment. When a device is not zeroed, it causes a continuous error, which is more obvious for smaller measured values than larger ones.

Calibration (systematic) – Calibration of an instrument should be checked before collecting data whenever possible. If a calibration standard is not available, the accuracy of the instrument should be validated by comparing it to another instrument of comparable precision or by examining the manufacturer's technical data. Because calibration errors are typically linear (measured as a proportion of the total scale reading), greater values result in larger absolute inaccuracies.

Physical variations (random) - Obtaining as many measurements as possible throughout the widest possible range is typically a good idea. When done on a regular basis, it reveals discrepancies that might otherwise go unnoticed. These distinctions may need more examination, or they may be blended to get an average value.

Parallax (systematic or random) – This mistake can occur when there is a gap between the measuring scale and the indicator used to collect a measurement. If the observer's eye is not correctly aligned with the pointer and scale, the reading may be excessively high or low (some analogue metres have mirrors to help with this alignment).

Instrument drift (systematic) - Most electronic equipment' measurements drift over time. Although the amount of drift is normally minor, this source of error might be considerable at times.

Systematic lag time and hysteresis - Some measuring equipment require time to attain equilibrium and taking a measurement before the instrument is stable will result in an unreasonably high or low value. A common example is taking temperature measurements with a thermometer that has not reached thermal equilibrium with its surroundings. An analogous phenomenon is hysteresis, which occurs when instrument readings lag and appear to have a "memory" effect when data is recorded consecutively moving up and down throughout a range of values. When a changing magnetic field is applied, hysteresis is mostly associated with magnetised materials. (webassign.net, 2011)

Conclusion

The wind tunnel experiments were a great way to reiterate the theoretical principles learned in class. Observing the effects of airflow around an aerofoil visually as well as graphically also helped in gaining additional insight into the mathematics behind aircraft aerodynamics.

Analysing the experiment results qualitatively as well as quantitatively introduced a sense of balance and conciseness to the assignment. This assignment also required a lot of critical thinking.

Bibliography

AEROLAB.com, 2021. *CLOSED CIRCUIT WIND TUNNELS.* [Online]
Available at: https://www.aerolab.com/aerolab-products/closed-circuit-wind-tunnels/
[Accessed November 2021].

AeroToolbox.com, 2017. *Aerodynamic Lift, Drag and Moment Coefficients.* [Online]
Available at: https://aerotoolbox.com/lift-drag-moment-coefficient/
[Accessed November 2021].

arc.uta.edu, 2014. *Low Speed Wind Tunnel.* [Online]
Available at: https://arc.uta.edu/facilities/lowspeed.htm
[Accessed November 2021].

DNW, 2021. *The Transonic Wind Tunnel Göttingen (TWG),* Braunschweig: German-Dutch
WInt Tunnels.

Dr. Hui Hu, 2021. AerE 344: Undergraduate Aerodynamics and Propulsion Laboratory,
Ames: Iowa State University.

empoweringpumps.com, 2021. *THEORY BITES: BOUNDARY LAYER (FLUID
MECHANICS).* [Online]
Available at: https://empoweringpumps.com/theory-bites-boundary-layer-fluid-mechanics/
[Accessed November 2021].

Encyclopaedia Britannica, 2018. *wind tunnel.* [Online]
Available at: https://www.britannica.com/technology/wind-tunnel
[Accessed November 2021].

eng.yale.edu, 2021. *MASON LAB B-7 WIND TUNNEL OPERATING INSTRUCTIONS.*
[Online]
Available at:
https://www.eng.yale.edu/metl/Wind%20Tunnel%20operation%20instructions11.pdf
[Accessed November 2021].

gla.ac.uk, n.d. *JWS732, aerodynamics and propulsion laboratory users' notes.* [Online]
Available at: https://www.gla.ac.uk/media/Media_392738_smxx.pdf
[Accessed November 2021].

grc.nasa.gov, 2021. *Blowdown WInd Tunnel.* [Online]
Available at: https://www.grc.nasa.gov/www/k-12/airplane/tunblow.html
[Accessed November 2021].

grc.nasa.gov, 2021. *Boundary Layer.* [Online]
Available at: https://www.grc.nasa.gov/www/k-12/airplane/boundlay.html
[Accessed November 2021].

grc.nasa.gov, 2021. *Open Return Wind Tunnel.* [Online]
Available at: https://www.grc.nasa.gov/www/k-12/airplane/tunoret.html
[Accessed November 2021].

infoWERK, n.d. *Aerodynamis AF101 to AF109,*
https://www.infowerk.systems/data.cfm?vpath=ma-wartbare-inhalte/ma-
downloads/factsheets-teaching-equipment/af101-109_1015pdf: infoWERK Medien &
Technik GmbH.

Laurence K. Loftin, J., 2012. *High-Lift Systems.* [Online]
Available at: https://history.nasa.gov/SP-468/ch10-5.htm
[Accessed November 2021].

Ria & Kartik, 2021. *High Lift Devices And Their Purpose.* [Online]
Available at: https://www.aviatorsbuzz.com/techtalks/high-lift-devices-and-their-purpose/
[Accessed November 2021].

Sreeja Sreekumar PPT, 2021. Aero Tech II(Fall 2021) Wind Tunnel Lab. s.l.:s.n.

TecQuipment LTD., 2021. *subsonic Wind Tunnel 600 mm,* Nottingham: TecQuipment LTD..

TecQuipment, 2021. *Continuous Supersonic Wind Tunnel,* Nottingham: TecQuipment.

vlab.amrita.edu, 2015. *1. Wind Tunnel Fundamentals.* [Online]
Available at: https://vlab.amrita.edu/?sub=77&brch=297&sim=1700&cnt=3610
[Accessed November 2021].

webassign.net, 2011. *Measurements and Error Analysis.* [Online]
Available at:
https://www.webassign.net/question_assets/unccolphysmechl1/measurements/manual.html
[Accessed November 2021].